Tick Tock Time

R. Hammond

Illustrations by Elena Napoli

ROMEO HOTEL PUBLISHING

Tick Tock, TIME is published under Romeo-Hotel Publishing, LLC.

ROMEO HOTEL
PUBLISHING

Virginia Beach, Virginia

In digital and print distribution.

www.rhammond.org

Library of Congress Registration Number: 2020923971

Hardcover & eBook

ISBN: 978-1-7349885-1-2
ISBN: 978-1-7349885-3-6

Author and Design by R. Hammond

Illustrations and Design by Elena Napoli

Florence, Italy

First Edition

For educational and entertainment purposes for our children of the stars.

This book can be ordered in bulk through Romeo-Hotel Publishing at www.rhammond.org

“Not until we have taken a look into the future shall we be strong and bold enough to investigate our past honestly and impartially.”

— Erich von Däniken, *Chariots of The Gods*

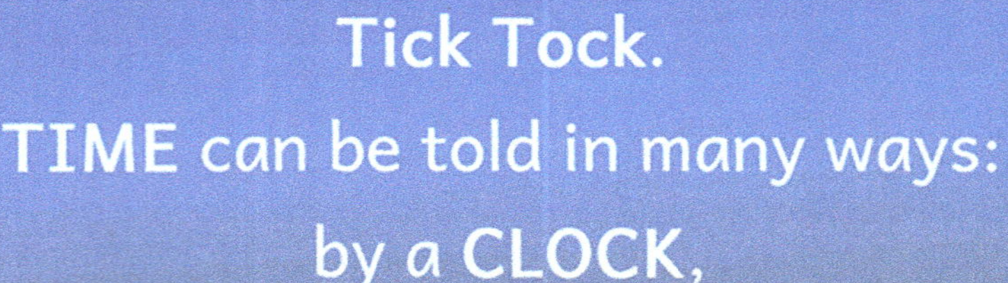

Tick Tock.
TIME can be told in many ways:
by a **CLOCK**,

by the STARS,

by the rise and fall of the **SUN**,

by the turning
of the SEASONS,

by the movement of the **TIDES**,

by the strength of the WIND,

by the change to EARTH,
PANGEA

by the growth
of **HUMANS**,

ANIMALS,

and of a **PLANT**,

by the finding of **FOSSILS**,

by the step of the **BEAT**,

by the types of **TRAVEL**,
near and far, by **SEA**,

by AIR,

by LAND,

to SPACE,

by the steel in *a* **FIELD**,

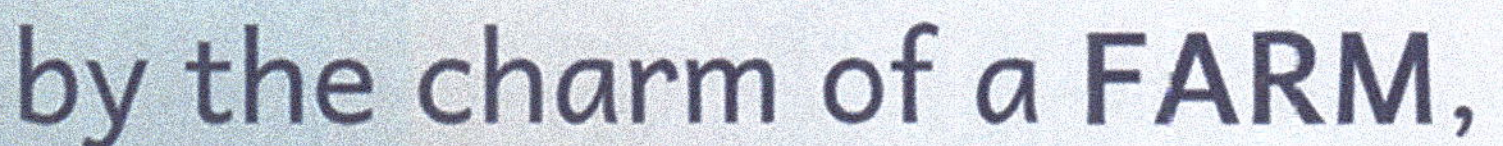

by the charm of a **FARM**,

by the strife
in LIFE,

by the hum
of the **HOME**,

by the bustle
in the **OFFICE**,

by the tone of the PHONE,

by the riddle
in the HOSPITAL,

and where
you LIVE,

second-to-second,
minute-to-minute,
hour-to-hour,
day-to-day,

week-to-week,
month-to-month,
year-to-year,
to-year,
to-year...

Tick Tock, TIME

Prof. Niko

HOROLOGIST

CPSIA information can be obtained
at www.ICGtesting.com
Printed in the USA
BVHW020133200321
603098BV00011B/839

9 781734 988512